Conquering Self Defense

How to Protect Yourself and Those You Love

Conquering Life Series

SOCIEDAD de los CONQUISTADORES

Edward A. Ipser, Jr.

Conquering Self Defense: How to Protect Yourself and Those You Love by Edward A. Ipser, Jr.

Conquering Life Series

ISBN 978-1-883661-09-0

Published July 12, 2020

Edited 1/10/24 4:44:00 PM

For Samantha, Antonia, Katrina, Aaron, Alan, and Ada.

Special thanks to ArLyne Diamond, Scott Flint, Wallace Gardner, Dennis Kai Ta, Declan McCullagh, Patrick Peterson, Dave Truslow, and Surjatini Widjojo.

Table of Contents

Introduction

I've had a lifelong interest in military history, wargames, and target shooting, I boxed in high school, and I did a self-defense startup with a retired police officer. All of this has led me to take an interest in self-defense. But becoming a husband and father provided the biggest motivation to get serious about it.

Nevertheless, I've managed to pass middle age without ever getting into a serious fight. Nor have I ever been a victim of violent crime.

However, I have been a victim of property crime on several occasions. I've lost more than one bicycle to thieves. I had a necklace snatched from my neck on a public bus. My wife and I were confronted by a belligerent drunk on the street. Our apartment was burglarized. I caught a pickpocket trying to steal my palm pilot on a subway platform. And I confronted a thief trying to steal my car in the middle of the night. Things could easily have gone differently on any of those occasions.

Conquering Self Defense

I wanted to master self-defense, so I called on the conquistador and asked him to enlighten me. This is what I learned.

History

Self-defense is not just ancient; it is inherent to the biology of life itself. Life has advanced through a process of natural selection that includes predators, prey, territorial defense, and, therefore, mortal conflict.

This fact of life was carried forward into human society in the form of the martial arts. In tribal and agrarian societies every able-bodied man was a soldier and social justice was a matter of personal, clan, and tribal honor.

The martial arts, as we know them today, derive from this tradition. They are, essentially, mock combat: practice and training for human aggression. Most modern sports originate from this tradition including, obviously, wrestling, karate, and boxing, but also team sports like football and tests of strength, endurance, and skill such as javelin throwing, and running.

Along with the refinement of human aggression, the martial arts developed an ethical system that is characterized by two main principles: sportsmanship and self-discipline.

Games of mock combat require and teach sportsmanship. Rules are developed to constrain and channel aggression in ways that limit injury, but which also develop interpersonal skills. For example, when a game or match is finished the participants shake hands in a friendly gesture, something that can be especially difficult for the loser.

Self-discipline comes into play in several ways. Training to develop strength and skill requires self-discipline. And practicing sportsmanship in the heat of a game or match requires self-discipline. But especially for those trained in violent martial arts, self-discipline is required to properly constrain violent aggression to those real-world situations in which it is morally justified. The more skilled you become about how to injure or kill someone, the more careful you must be in any confrontation.

With civilization, however, it became possible for people to delegate justice and security to civil institutions. And classes, castes, and professions developed who specialized in martial arts, warfare, and policing to the exclusion of others.

In many societies only nobles were permitted to carry arms. As late as the seventeenth century, the carrying of arms, openly or concealed, was the mark of a gentleman. An armed gentleman was considered a step above the common man. From time to time, there was a revival of the actual practice of carrying arms, usually in response to a rise in anarchy and crime. In the late nineteenth and early twentieth centuries, European nobility renewed their interest not only in arms but in the martial arts as gangs of notorious professional thieves and robbers began preying upon them. If the criminal was proficient in violence, then a gentleman should have no qualms either. People developed skills with walking sticks, knives, truncheon, swordstick, and pistol.

Martial arts became increasingly oriented toward the exceptional situation of defending oneself against immediate crime which, however common, was distinguished from justice and punishment administered by formal institutions of society.

The modern state came into being in the sixteenth and seventeenth centuries. The defining concept of the state, as understood by political philosophers, was that it held a monopoly on the legitimate use of violence.

Nevertheless, virtually every society left intact the individual right of self-defense.

The Catholic Church's teaching, for example, is that violence is justified in self-defense up to and including taking the life of an assailant. Defensive violence is not merely a natural right, but also a grave duty. Saint Thomas Aquinas explained that "if a man, in self-defense, uses more than necessary violence, it will be unlawful: whereas if he repels force with moderation his defense will be lawful."

The role of martial arts thus became bifurcated between, on the one hand, practice for war conducted by states against one another and, on the other hand, carefully constrained situations in which a victim had no other recourse than to fight back against a criminal assailant.

It is tempting to distinguish civilian self-defense from military warfare by the nature and intensity of the conflict. Civilians are not at war, after all. But this is illusory. An infantryman spends most of his time training, preparing, and waiting. Even in a war zone, it is hours of boredom punctuated by minutes of violent terror. And counterinsurgency war is even more like self-defense because there is no front line; an attack may come at any time from any direction. War and self-defense are not so different after all.

Courts have consistently ruled that the police have no legal duty to come to your aid. So, while it is prudent to delegate our security to others in ordinary situations we must, nevertheless, reserve to ourselves the right to self-defense in extraordinary situations. As the saying goes, "when seconds count the police are just minutes away."

Definition

Self-defense is the rightful use of violence to defend oneself or others against wrongful harm. The goal of self-defense is the protection of personal health, freedom, and property. We don't use self-defense to impose our will on others but to prevent others from violently imposing their will on us.

We each have an inherent, natural right to life, liberty, and property and, thus, an inherent and natural right to defend these. This right does not depend the concurrence of others, not even on the enactment of law, because laws that contradict natural rights are inherently unjust and immoral. You can't give away your right to self-defense and nobody can take it away from you. The statesman and philosopher Cicero said, "if our lives are endangered by plots or violence or armed robbers or enemies, any and every method of protecting ourselves is morally right."

Legitimate self-defense has three criteria: 1) someone poses an imminent and unlawful threat of death or great bodily harm. An imminent danger is an immediate or present threat happening right before you and cannot relate to something that may or may not happen in the future. 2) The immediate use of force is necessary to avoid that danger. And 3) you use no more force than necessary. Someone who has been threatened or harmed by an assailant in the past is justified in acting more quickly or taking greater self-defense measures against that assailant.

If your situation meets those criteria then you are entitled to use deadly force, force likely to cause death or great bodily harm. One corollary is that the right to self-defense ends when the threat ends. So, for example, if you confront someone in your home and he turns his back to leave then he is no longer a threat of serious injury or death, and you have no right to attack or pursue him.

While no legal definition of self-defense can override the natural right to self-defense, common and statutory law can help clarify ambiguities and regulate its exercise. Practically speaking, the question boils down to whether a person who exercises what he believes to be self-defense will be convicted of assault or homicide. This will vary with the jurisdiction, the unique circumstances of the event, and the attitudes of the police, prosecutors, judge, and jury evaluating the matter. Each person involved in judging your decision will be asking himself, "if I had been in his shoes would I have done the same thing?" This is referred to as reasonableness and it is very subjective. Note that a belief can be reasonable even if it is not correct. In determining if a belief is reasonable, the judge and jury will look not only at the facts of the situation as they know them at the time of the trial but also at how those facts appeared to you when you exercised self-defense.

Ultimately, though, as the saying goes, "it's better to be tried by twelve than carried by six."

For the most part, property crimes such as robbery do not merit a deadly force response. But not always. If great bodily harm is threatened in the course of a property crime, then deadly force is justified. It may be prudent to cooperate with an assailant, if only temporarily, but you are under no moral or legal obligation to cooperate in the commission of a crime against you. And if you reasonably fear that your noncooperation will lead to a violent attack, you are justified in using deadly force. If, for example, someone threatens you with "your money or your life" you are entitled to defend your life. Abduction, kidnapping, and carjacking all pose such threats. And, in the unlikely event that the property crime represents a threat to your survival, such as stealing your family's last loaf of bread, that may be a situation where deadly force is justifiable.

The defense of property may also amount to self-defense when attacks on property are designed to undermine civic order. Rioting and looting are often directed not simply at the property they destroy but law and order itself. The preference for leaving the defense of property to law enforcement and the judicial system presupposes that these are operating in good order. When civil society is undermined, violence becomes the only means of resolving disputes. Self-defense against marauders is justified because they threaten the very system that would otherwise deal with them as an alternative to self-defense. Political commentator Andrew Sullivan said, "Any civil authority that permits, condones or dismisses violence, looting and mayhem in the streets disqualifies itself from any legitimacy."

In most jurisdictions, if someone invades your home, they are presumed to be a threat not just to your property but to your life. This is sometimes codified as 'castle doctrine' law. In some legal jurisdictions, you have a duty to retreat, if possible, even in your own home, but especially outside it. In other jurisdictions with 'stand your ground' laws, you have no duty to retreat even in public.

But, as a general rule, don't fight for anything less than a human life. There are many lesser threats which the law deals with more subtly: hurtful words, threats in the future, attacks on your honor, pugilistic fist fights, or simple violations of property. All of these are in a gray area where self-defense is generally not reasonable.

We are all, each of us, primarily responsible for our own personal safety. Additionally, we might be responsible for the safety of others either in our care, such as children, those with whom we have a relationship, or even, in some cases, complete strangers. If we delegate this responsibility to another, even to the state, we assume the risk that they might not fulfill that responsibility leaving us exposed to violent crime.

Threats

The incidence and nature of violent crime varies widely from country to country, from state to state, from city to city, and even from neighborhood to neighborhood. Some violent crimes, like robbery, are commonplace while others, such as terrorism, are dramatic but rare.

It's important to distinguish an assailant who is bad tempered and lacks self-control from a calculating criminal. Both are dangerous but the criminal is more likely to use cunning to catch us off guard and at a disadvantage.

Assault and Battery

Assault is sometimes defined as any intentional act that causes another person to fear that he is about to suffer physical harm. Historically, battery was the completion of assault. But the distinction is often ignored in common usage either by using the phrase "assault and battery" or by simply treating assault as including both the threat and the act. Aggravated assault is any assault with a deadly weapon (fists, a knife, a gun, or even an automobile) or against a vulnerable person with the intent to cause death or an injury that requires medical attention.

Burglary and Robbery

Burglary is any unlawful entry into a home, car, or other location, usually by stealth, for the purpose of theft or other property crimes. Robbery, on the other hand, is unlawfully taking, or threatening to take, property by violence. Unlike burglary, robbery necessarily entails the physical victimization of the property owner or caretaker. The most common form of robbery is mugging, or street robbery, in which the assailant attacks, or threatens to attack, his victim in order to steal property on his person, such as a wallet, purse, watch, or jewelry. A store or bank robbery, in which the assailant attacks, or threatens to attack, an employee or proprietor, is another common example.

Home Invasion

Home invasion is the forceful entry into an occupied home with the intent to commit a violent crime such as robbery, assault, rape, kidnapping, or murder. Most criminals prefer to burglarize an unoccupied home and will go to great lengths to ensure that the home is unoccupied before entry. So, a deliberate home invasion, when it occurs, is a very serious matter that suggests deadly intentions.

Carjacking

Carjacking is a robbery in which the assailant takes a car by force from the occupant. Very often carjacking results in kidnapping since the driver or another occupant, sometimes a child, may still be in the car. Carjacking may be merely a means to a more dangerous end such as kidnaping or rape.

Rape

Rape, or sexual assault, is most typically the act of a man forcing a victim, usually a woman, into sexual intercourse, but it also includes other similar sexual violence. Rape may result in bodily injury, an unwanted pregnancy, the transmission of deadly disease, or psychological trauma.

Kidnapping

Kidnapping, or false imprisonment, entails the forceful abduction, transportation, and confinement of a person against his will. The motive for kidnapping may be rape or ransom. In addition to the assault involved in the act of kidnapping, victims very often become rape and murder victims.

Murder

Murder is the intentional, unjustified killing of a person. Manslaughter, on the other hand, is an unintentional killing either by negligence or diminished capacity.

Lawful self-defense that results in the death of an assailant is not murder, it is a justified killing. But an error in judgement by the victim can result in a charge of manslaughter or murder.

Terrorism

Terrorism, bombing and mass shootings, is a broad term that typically entails violence for political purposes, as for example when victims are killed in order to instill fear in a broader population or to bring attention to some political cause.

Awareness

Awareness is perception and knowledge of the potential dangers we face in any given situation. It is the first step towards self-defense.

Be aware, first of all, of the very real danger that criminals present. Don't assume everyone else is as peaceful as you. Firearms instructor Jeff Cooper said, "Some people prey upon other people. Whether we like it or not, this is one of the facts of life." The problem is that we become habituated to our tranquil environment and overly trustful of others around us. Certainly, most situations are perfectly safe. But whether you are at home, at work, walking or driving around town, or traveling, there are people out there who may prey on you, especially if they think you are unaware. You can become a victim of crime at any time, in any place. But if you are aware and know how to respond then you will be better prepared to do what is necessary if you are ever attacked.

Be aware of your own sense of fear. Thanks to our biological history, we all possess a natural and intuitive fear of dangerous people and situations. This natural skill should not be ignored or rationalized away but, instead, respected and refined.

Avoid high crime areas. It's a plain fact of life that some places are more dangerous than others, often by orders of magnitude. For example, avoid dangerous parties, bars, and clubs with a reputation for trouble.

Stay in well-lit public areas at night and avoid dark secluded places. For example, a parking garage may be busy and well-lit during the day when you park but dangerously dark and secluded when you leave.

Avoid drawing attention to yourself. Be aware of how your dress or behavior may encourage criminals to believe that you are an easy or worthwhile target. Don't wear expensive jewelry in public. Dress in a way that blends into your environment.

Most crimes happen at night, so it is safer to shop or run errands during the day. But if you are out at night then find a parking spot close to the store and under good lighting to minimize and carry a flashlight. Even if you are out alone you can still avoid being alone by walking with strangers.

Note the exits in any room you enter. Sit with your back to the wall and facing the door at restaurants or other public establishments whenever possible.

To avoid being a victim, don't put yourself in risky situations. If you find yourself in a risky situation, remove yourself from it.

Alertness

Criminals seek advantages over their victims including, especially, surprise. To avoid becoming a victim of crime, avoid being surprised. To avoid being surprised, be alert and always ready.

Being alert means being aware of your surroundings, especially what is behind you. It means knowing who is in your immediate vicinity and what is happening around you, when things are out of place, and when people are giving you undue attention. It means not letting yourself get distracted by conversations and sight-seeing. The cellphone is the worst enemy of alertness, people become engrossed in it and lose all awareness of their surroundings. Be alert to everything 360 degrees around you, particularly in your path ahead.

Alertness is not paranoia. Marine Corps ace Joseph Foss said, "looking doesn't cost anything and it is a healthy habit." Make a habit, even a game of it. Keep your eyes moving. Look at everything in your field of vision. Know who is knocking before you open the door.

Federal law enforcement officer Gary Quesenberry says, "You can't fight what you don't see coming." According to Federal law enforcement officer Gary Quesenberry, situational awareness is more than just keeping your head up and your eyes open; "It's learning how to read the environment and people. How are they carrying themselves? Are they doing something that catches your eye? Look for activity where the person is actively scanning the area and their behavior is not calm. When most people are walking from their car to the store, or when exiting a store, they are calm and relaxed. But when you see someone jittery and nervous or agitated, and their eyes are scanning around, that's a sign."

Self-defense instructor Steve Mittman advises, "Scan the entire area when approaching your vehicle. Listen for running cars. Look for parked cars with their lights on. Watch for people sitting in their cars and people walking or standing between cars. This is all unusual behavior and could be a sign that the criminal is waiting in their car or walking around the parking lot for their next victim."

If you arrive at your home or business and find signs of forced entry, don't enter. The intruder may still be inside. Instead, call the police.

There is a saying, "when two tigers fight, one is killed and the other is mortally wounded." Criminals are looking for easy targets that don't present a risk to their getting away safely. They generally avoid confident and alert victims. A criminal will not want to risk getting hurt, or even inconvenienced, when there are so many easier victims out there. Alertness makes you a hard target and will discourage criminal interest.

Maintain boundaries with strangers, especially when no one else around. While one and a half feet is typical for a friendly conversation, and two to six feet in social situations, keep suspicious strangers at 25 feet or more.

Assailants will often indicate their intent by their behaviors. They will observe and interview their victims. Their goal is to determine if you are a suitable victim. An assailant may scan for inattentive people who look like easy victims. Or he may approach you and ask, "can I ask you a question?", "what time is it?", "can I have a light?", or say, "you dropped this!" He may shock you with a confrontation designed to test your reaction, meek or firm. Or he may attempt forced teaming with directing commands about what you and he must do together.

Be alert to people milling about aimlessly. Take notice when you see someone make an abrupt change of direction or reorientation that doesn't make sense, especially if it seems like an orientation on you or deliberate movement towards you. Watch for coordinated movement or nonverbal communication among multiple people oriented toward you.

Robbers use many tricks to distract you and make you vulnerable. For example, they may place money on your windshield wiper that you only see after unlocking and getting into your car forcing you to get out again. This can be a setup for a carjacking or a robbery.

If someone draws your suspicion, watch his hands. Are his hands inside his clothing or is he constantly touching his waistline or adjusting his clothing? This could be a sign of a hidden weapon.

When an assailant is ready to attack, he will close the distance with speed and/or stealth. Victims often say, afterwards, that the assailant "came out of nowhere." He may attempt to fix you in place by blocking your path or, if there are multiple assailants, one may attempt to distract you while the other closes for the attack or one may herd you into others.

An assailant may demonstrate several pre-attack indicators. To begin with, he will move within striking distance. He may look around to see who is watching. He may cover his face with a hand to obscure his intentions. He may groom his hair or give a 'felony yawn.' He may begin shrugging and flexing his shoulders. He may quicken the pace of the conversation. Or he may dehumanize you with insults or begin speaking incoherently.

As his adrenaline kicks in he may sway from side to side, rock on the balls of his feet, pace back and forth, increase his rate of respiration, blink faster, stare, or his hands and knees may begin shaking. He may begin shaking his head 'yes' as he holds an internal dialog to pump himself up.

He may shift into a fighting stance with his weight to his strong back leg or crouch slightly as if to pounce. He may take off his shirt, clench his hands, ball his hand into a fist, pump his fists, bring his hands up into a fighting stance, glance at possible targets, or load up to launch a fist. Or he may move his hands to his back or into a pocket to draw a weapon.

Be especially careful not to let a single assailant draw your entire attention, he may have confederates waiting for an opportunity to strike you when you are distracted.

Decisiveness

When danger presents itself, it is time to set aside what you were doing and focus on it. We can easily become locked into completing a task. People are often victimized at automatic teller machines because their attention to withdrawing cash makes them inattentive to potential dangers around them. Even when danger presents itself there is a powerful temptation to complete the task before dealing with it. If you notice a threat, you should orient on it and start thinking about how you will respond to an attack.

If you suspect an attack is imminent, then you have several options: evasion, de-escalation, and sounding the alarm. Regardless of what you decide, you must be decisive.

If you have the time and space, look for an exit from the situation. Finding a place where there are potential witnesses is usually sufficient.

If you cannot find an escape, you might try to de-escalate the situation by making conversation, using flattery, or even lying. You are under no moral obligation to be truthful to an assailant. You are perfectly justified in deceiving him if that will allow you to escape or otherwise avoid becoming a victim. Occupy his attention or confuse him to buy time.

Offer your potential assailant a face-saving exit with sincere humility. Maybe you just rubbed him the wrong way.

In some situations, a gentle, physical touch can be used to communicate friendliness.

Use a firm voice to give a sharp command. But a command backed up by a weapon in hand is more persuasive.

Throw a decoy (or real) wallet on the ground and run.

Remember that there is no moral or legal requirement to cooperate with an assailant. In many cases, cooperation can be dangerous. For example, in an abduction, the first goal of the assailant will be to quietly move you to a place, or to place you in a condition, where he has full control. Your choice may be to fight now or die later.

If you can't evade and you can't deescalate, but you believe that assistance might be within earshot, then try sounding the alarm either by yelling or screaming or using a whistle or horn.

Now is the time to think about what you are prepared to do under what circumstances, how you plan to respond to potential threats. Make a habit of imagining various hypothetical scenarios. "What would I do if this person I am currently interacting with suddenly demonstrated aggression?" "What would I do if I found my path blocked by an assailant?" "What would I do if someone invaded my home in the middle of the night?"

Once you come under attack, the time to think about acting is past. The time to act decisively has arrived.

Aggressiveness

When all else fails, if an attack is imminent, then the time for violent action has arrived. You must make a decisive, instant transition from amicability to aggression. Because the best defense is a good offense you must attack the attacker, striking first if possible.

Catatonic shock or stupor, a state of immobility, is a natural reaction to being overstimulated as, for example, when you unexpectedly fall victim to violence. You can minimize the risk of catatonic shock by heightening your natural level of alertness and by developing a mentality for combat. Marine General James Mattis said, “Be polite, be professional, but have a plan to kill everybody you meet.”

If you are sufficiently aggressive it may be possible to surprise your assailant by turning the tables on him. Criminals expect their victims to be afraid and compliant, not to attack them. That’s how they choose their victims in the first place.

This is an unnatural idea for most people nurtured on belief that the good guy must give the bad guy a free swing. However, depending on your situation and your strength and skill relative to the assailant, such a passive approach could invite the worst outcome.

To make this transition successfully it helps to set a mental trigger. Establish go/no go criteria. Decide, ahead of time, what you are willing to do in what situations. If you've already thought through your 'what ifs' then this simply entails transitioning into your prepared response.

When you trigger your aggressive response, move with speed. Fights are blindingly quick. Keep your cool and keep it simple. Act as you've prepared and trained. Cultivate and arouse indignation and righteous anger against the unjust threat to your personal security and welfare.

When we are facing a threat of violence, we get a rush of chemicals to prepare us for combat. Sudden stress triggers the release of adrenaline and endorphins that help our body change from its normal condition to a state of readiness, a level of awareness that helps us decide on a reaction to the crisis and reallocate our body's resources. We experience tunnel vision to focus on the threat. Our heart rate goes up, we get the 'shakes', and begin rapid, shallow breathing. We experience auditory exclusion, not hearing much of what goes on around us. We experience an altered perception of time.

Air Force Colonel John Boyd developed a decision-making model called the OODA loop. When we face an assailant, we first observe, discerning our adversary's intent, then orient, then decide, and then act. At the same time, we want to obscure our own intentions as best as possible. Think about a feinting boxer or a juking halfback. Boyd's crucial insight was that we can gain an advantage by going through this loop faster than our adversary so that we gain a step ahead of him in our actions. This is particularly important in a self-defense situation because the assailant already chose when, where, and how to attack.

Once you have achieved an advantage over your assailant, maintain the momentum of your counterattack until he is stopped, whatever it takes. If you catch your attacker off guard, you will have a momentary advantage. Do not surrender that advantage.

Be ruthless in your counterattack. If deadly force is justified, use it. Fight hard, fight dirty. Self-defense is not a sport; and it is never a fair fight. There is no tapping out in self-defense. If self-defense is justified, then deadly force is justified. If deadly force is not justified, then self-defense is not justified.

Your goal is to stop the assailant. To accomplish this, you must inflict pain and damage his function. The death of the assailant is never the objective, at worst it is a means to an end: stopping him.

Expect to get hurt but don't worry about it. Focus, instead, on winning and surviving. You will be surprised at how your adrenaline helps you to ignore the pain of injury while you are in a fight for your life.

But keep your ego in check and know when to stop or retreat. Never take revenge. When the assailant stops, whether by surrendering, fleeing, or falling unconscious, you must stop your counterattack. If your counterattack opens an opportunity to escape, you can take it but don't forget about those you might be protecting. Remember, it's not your job to prevent crime, generally, or to punish criminals. That's the job of the police. Your job is simply to avoid becoming a victim of crime and to protect your family.

Technology

Home security systems include motion sensors, contact sensors, and cameras usually connected to a monitoring system in the home and, often, a central monitoring office or a smartphone. Home security systems are mainly designed to prevent burglary, but they can also provide a critical early warning of a home invasion in the middle of the night.

There was a time when home security systems were only for the wealthy. But modern technology has brought these into the price range of almost anyone. In addition to deterring criminals and providing you with warning of a break-in, installing a home security system can reduce your home insurance costs.

In addition, you can install lighting around your house, either solar powered and always on at night or else connected to a motion sensor.

Senior alert systems, devices worn around the neck that allow you to summon help, have been around for decades. But the ubiquitous smartphone can serve as a tool for security in many additional ways. First, it is a mobile phone that you can use to summon assistance when you are away from a landline phone as, for example, if you are out walking or driving. Even at home or at an office a cellphone is more likely to be at your fingertips than a landline phone.

Because smartphones include a camera, you can take photographs and send them as a text message to a friend or family member to let them know where you are, who you are with, and the situation you are in. For example, you can text the license plate of a taxi or ride share car before you get into it. And because smartphones include GPS positioning, they can be used to find you in an emergency.

There are smartphone phone apps that provide an extra measure of security. For example, if you are traveling in a taxi or ride share and you feel uncomfortable with the driver you can use a 'dead man switch' that alerts the police or someone else if you release the button.

Natural Weapons

Weapons are the instruments of self-defense. A natural weapon is anything that you were born with including your arms, legs, hands, elbows, knees, and even your head. This is sometimes called unarmed self-defense, but you should, instead, think of your fists as natural weapons.

Your natural weapons should be directed against the vital targets of your assailant. A vital target is one that is particularly sensitive or crucial. But because natural weapons are sensitive and fragile it is important to match them against appropriate vital targets.

Some targets are more sensitive than others. For example, a punch to the nose can be very painful but it will not necessarily stop an assailant; someone with a broken nose can easily fight on. Injuring muscles is painful and discouraging but not usually debilitating. But these may be enough to discourage an assailant who is looking for an easy victim.

On the other hand, a blow to the neck or to the jaw can cause a blackout. An eye gouge, in addition to being very painful, will partially blind an assailant. And a crushed windpipe will cause suffocation.

Vital targets in the front of the body include the top center of the skull, eyes, temples, ears, windpipe, carotid arteries and jugular veins, heart, solar plexus, groin, knees, shins, and instep.

Vital targets on the rear of the body include the base of the skull, the spine, the back of the knees, and Achilles tendons.

When selecting a target, consider:

- Criticality: how vital is the target?
- Accessibility: How easily can you strike the target with an appropriate natural weapon?
- Recognizability: How easily can you find the target?
- Vulnerability: How much force is needed to affect the target?

Generally speaking, you should direct natural weapons against vital targets appropriate to their relative strength. For example, a knee is stronger than a pelvis, a shin is stronger than a hamstring, an elbow is stronger than a chin, a palm is stronger than a nose, and a finger is stronger than an eyeball.

If someone is moving aggressively into striking distance and you fear that he might attack, but you are not certain enough to strike first, then put up your hands in front of you with your palms open and facing him. This will help you guard against a sucker punch without overtly displaying a hostile intent on your part. Bend your elbows out to guard against a wide haymaker swing.

If the assailant strikes, then quickly move into a turtle defense with your arms converging at your head to ward off the initial blows. But while it is very tempting to remain in a defensive position, reliance on this as a primary tactic will simply leave your assailant free to continue his attack until he finds an opening in your defense. Instead, use a defensive posture as a reflex against a surprise attack and transition immediately into a counterattack.

Effective use of natural weapons requires a good fighting stance. A good fighting stance provides a stable, balanced platform for striking with your natural weapons as well as flexibility and mobility. To get into a good fighting stance, spread your feet roughly shoulders width. Extend your dominant foot slightly behind you. Keep your knees bent slightly, not locked. And stay on the balls of your feet.

Your hands are your most precise natural weapons. A conventional punch is almost instinctual, but it risks damaging the hands, perhaps even breaking a bone. A hammer fist is simply a downward blow of the clenched fist. A reverse punch is a swing of the back of your hand. Open handed strikes, using your palm or the knife edge of your hand, can also be effective against weaker targets.

If you've watched mixed martial arts, or cage fighting, you'll notice that much of the fighting consists of basic boxing techniques. Simple punches to the head or torso are natural and can be effective. But as unrestricted as MMA appears to be, it is still a sport designed to limit injury to the fighters so there are many effective techniques that are prohibited.

For example, you can attack your assailant's eyes with a finger stab. Extend your index and middle finger straight and fold back your thumb, ring, and little fingers. Aim for the eye socket and stab as hard as you can. Or jab or gouge your thumb into your assailant's eye. If you are close enough to touch him with the palm of your hand, you are close enough to gouge his eyes.

Strike your assailant in the throat with a panther strike. To form your hand into a panther strike, fold your fingers back at the second knuckle.

Use the palm of your hand to strike your assailant in the nose. Turn your hand up at a 90% angle and then thrust your palm into your assailant's nostrils driving his nose up.

Fold your arm with your hand to your chest and swing your elbow like a short club, either vertically or horizontally. Your elbow is much stronger than your fist; it can do great damage to your assailant by striking him in the head, neck, or upper torso.

Use your knee to strike your assailant in the groin, legs or, if he is bent over, even his head.

Use your head to butt your assailant's head, preferably using a stronger portion of your head, such as your crown, against a weaker portion of his, such as his nose or chin.

Kicking is not a very natural method of attack, it requires training and practice, but you can weaken an assailant with simple, straight kicks to his thighs or calves. And if you find yourself on the ground you can lay on your back and strike out at him with kicks.

You can blow out your assailant's knee by kicking or stomping it from the side or just above the kneecap.

While you should not rely on a defensive posture in self-defense, you may have an opportunity for an offensive block. An offensive block is a strike against your assailant's attacking arms or legs, typically intercepting them at a vulnerable place. For example, if your opponent is throwing a wide swing, you can thrust the bone of your forearm against the softer muscle of his arm. Likewise, if he is throwing a kick you can counter with a kick that throws your shin against his calf, thigh, or foot.

It is said that every fight eventually goes to the ground, at which point it becomes a wrestling match. In MMA, a fighter skilled at ground fighting will often look for a way to take down his opponent. But self-defense is different. You certainly don't want to let an assailant pull you to the ground. Once you go to ground you lose mobility and give up the option to run away. Worse still, you become more vulnerable to other assailants. So, avoid going to ground unless you are very confident that you are facing only one assailant and that you can win a ground fight.

If you are facing multiple assailants then tactics become crucial. You should always be looking to escape, but especially when you are facing multiple assailants. In dealing with multiple assailants, tunnel vision is your enemy. Keep your peripheral attention active. Avoid the temptation to focus on the assailant immediately in front of you. Instead, use the person in front of you to hide from others behind him and as a shield. Maneuver to take on your assailants one at a time so that they are in a column, one behind the other, rather than in line, all facing you together. So long as one assailant remains in the fight, you should be especially ruthless towards every one of your assailants, even those who have been knocked to the ground. Kick them when they are down to keep them down.

Nonlethal Weapons

Nonlethal weapons are intended to force compliance by inflicting discomfort or pain without the risk of injury or death.

A powerful flashlight can temporarily blind an assailant at night allowing you to make an escape. A strobe light is disorienting which makes it difficult for him to concentrate on his attack.

Mace and pepper spray are nonlethal defensive sprays that, when successfully delivered to the assailant's eyes, cause pain and, perhaps, temporary blindness. However, even when successfully delivered on target, the effectiveness varies, and some people are better able to shake off the effects than others.

Mace is a trademarked product made from a pepper solution. It is classified as an irritant because the solution is typically 10% capsaicin or less. Chemical mace can be ineffective against attackers who were under the influence of alcohol, narcotics, or psychotics. Since many assailants are under these influences, chemical mace is of limited value for self-defense.

Pepper spray adds a lachrymatory agent that causes tearing when applied to the eyes. It will also cause pain and swelling in proportion to the capsaicin in the spray. Pepper sprays that are used for self-defense purposes usually contain 10% capsaicin, which is similar to Mace-branded products, but law enforcement pepper sprays can be up to 30% capsaicin and are classified as a riot control agent. It may be illegal to own law enforcement pepper sprays in some jurisdictions.

The reaction to pepper spray is almost immediate and the swelling, pain, and blindness can last for minutes. It only takes a small amount to create results and you don't need to get it directly in the eyes for it to work. Law enforcement pepper spray has the potential of getting into the throat, restricting air passageways, and raising blood pressure at the same time. This increases the chances of a heart attack or stroke, which could put you in legal jeopardy if you use the spray in a jurisdiction where ownership is restricted or in a situation that did not justify deadly force.

Bear spray is a weaker form of pepper spray made from the same peppers as all the other sprays that are on the market, typically having around 2% capsaicin. This product is a good option to use for jurisdictions where higher concentrations are not allowed. It may still be restricted, but because the concentration is relatively weak, there are fewer consequences to using the product.

Electroshock weapons, such as a stun gun or Taser deliver an electrical shock to the assailant. The stun gun works by pressing two electrodes into contact with the skin either directly or through light clothing. The Taser fires two barbs attached to wires. The prongs stab into the skin and then a shock is delivered through the wires running back to the weapons battery.

Blunt Weapons

There are a variety of blunt weapons designed to cause damage by means of blunt force trauma. Blunt weapons range from improvised weapons, such as an umbrella or a chair, to martial arts weapons, such as a baton or staff.

The first rule of using a weapon in self-defense is this: don't ever pick up a weapon that you are not willing to use. Don't wave a weapon as a warning, the assailant just might call your bluff and take it from you. Once you select a weapon, look for the right opportunity to use it, preferably by surprise.

Improvised weapons are all around us. Take a moment to survey your surroundings and ask yourself, for each item you see, "how could I use this as a weapon?" A pen can serve as a stabbing weapon, a phone receiver as a hammer, a chair as a shield and club, and an ash tray as a missile. Other items that are not ordinarily classified as weapons can be improvised as such. For example, a baseball bat or an aluminum flashlight.

Traditional martial arts weapons include the staff (bo), usually about 71 inches in length, the baton or club (tonfa), usually about 15 to 20 inches, and stick, usually about 18 to 36 inches long. Each of these weapons has its own style of fighting that can be learned but they are simple enough that, in a pinch, anyone can use them more effectively than their natural weapons.

A staff can be swung like a baseball bat with your hands at the base. But for most strikes, you should position your hands spread with the strong hand at the ½ and the weak hand at the ¼ positions. In that position you can jab your assailant's body with the long end to keep him at bay or punch his head with the longer end. You can punch him with the short end in a quick strike pushing with the weak hand and pulling with the strong hand and turning the body. Or you can cross punch the assailant's neck.

Likewise, the traditional wooden walking cane with a curved handle can serve as a very effective weapon. You can carry a walking cane through security checkpoints where other weapons can't go including airport security and government buildings. It is a violation of the American with Disabilities Act for anyone to take your walking cane from you and it is a violation of the HIPPA privacy rule for anyone to inquire about your health, medical history, infirmities, or reason for carrying the cane. Hidden in plain sight, your walking cane is a weapon that you can employ with an element of surprise because your assailants will see it as a sign of weakness or disability.

The blackjack or sap is a simple but effective blunt weapon. It consists of a weight (usually loose lead or a steel rod) in a sturdy sack (usually leather) sometimes with a strap for added control. Blackjacks fell out of favor in the mid-20th century and are illegal to carry for self-defense in many jurisdictions today. However, a leather coin purse can serve the same purpose more discretely.

The tactical whip is a relatively new blunt instrument. A tactical whip consists of a steel cable with a plastic coating fixed to a handle.

A slingshot or wrist-rocket that uses high-tension rubber and firing steel balls can be effective in limited situations. It is not practical to carry and deploy in close quarters, but it can be useful to ward off someone who is illegally on your property. A slingshot can be fatal at 10 feet and cause serious injury at 30 feet. Beyond that it can serve as a painful deterrent.

The legality of carrying a blunt instrument as a self-defense weapon varies widely. In some jurisdictions, the mere carrying of a blunt instrument for the purpose of self-defense makes it an illegal weapon. But a flashlight, walking cane, coin purse, or staff is less likely to arouse suspicion.

Edged Weapons

Edged weapons are designed to inflict damage primarily by causing bleeding, although some edged weapons may also cause blunt force trauma. Edged weapons include knives, tomahawks, axes, spears, and swords.

Edged weapons can be used either to slash or stab. Some edged weapons are designed primarily for one method. For example, the Roman short sword, the gladius, is a stabbing sword whereas the cavalry saber is a slashing sword.

For the most part, the primary edged weapon that you will use in self-defense is the knife. Knives can be either fixed blade or folding. Kitchen and hunting knives are fixed blade. Some fixed blade knives are specifically designed to be fighting knives, such as the Bowie, Kukri, and Ka-Bar knives. The ordinary Swiss Army knife is an example of a folding knife. These can be closed simply by pushing on the dull side of the blade, which means that they can easily fold when they are being used to stab.

Folding knives can be locking. Locking folding knives hold the blade in the extended position by a locking mechanism that must be manually released to close the blade. Locking blade knives are well suited to self-defense because when the blade is locked into position, the knife acts like a fixed blade knife.

Edged weapons require more training and practice than blunt weapons. And they do not immediately incapacitate an assailant; he may eventually succumb to his wounds but unless you cut an artery it will take time.

Additionally, knives have a bad social reputation as the weapon of assassins and hoodlums and thus may draw extra scrutiny from a prosecuting attorney and less sympathy from a jury.

Firearms

Firearms are deadly weapons and, arguably, the most effective for self-defense. Firearms inflict physical trauma damaging muscle, bone, and vital organs depending on where and how the bullet strikes, as well as bleeding.

Because firearms are deadly weapons, it is important to faithfully follow firearm safety rules. The most basic rules are:

1) Treat all guns as if they are loaded.
2) Never point a gun at anything you are not willing to destroy.
3) Keep your finger off the trigger until your sights are on the target and you have made the decision to shoot.
4) Be sure of your target and what is behind it.

Overpenetration is a serious risk when using a firearm inside a home or apartment. There is a real danger that a bullet fired will pass through one or more walls to strike an innocent bystander. You are responsible for every round you fire even in self-defense.

Great care must be taken with the storage of firearms in the home, especially where there are children. Firearms can be stored in a gun safe along with other valuables, but this makes them inaccessible in an emergency. There are some simple mechanical locking devices that attach to the gun, but they are not always easy to remove in an emergency.

Firearms are complex mechanical devices. And while there are many general similarities, each model is a unique mechanical design, and each individual firearm can be uniquely customized. So, the first step in using any firearm for self-defense is to master the mechanics of your specific firearm at the gun range and in training.

Firearms used for self-defense must be reliable, accurate, effective, have sufficient ammunition capacity, and be ergonomically comfortable. A reliable firearm is one that never fails, even when it is abused. An accurate firearm is one that can put rounds on target when it is correctly aimed. An effective firearm is one that is powerful enough to stop an assailant when you put rounds on target. The capacity of a firearm is the number of rounds it carries, usually in an attached magazine. And ergonomics is the subjective fit of the firearm to your body in such a way that it is easy to manipulate and fire.

Guns fire ammunition: usually called a cartridge or round, or commonly, but inaccurately, a bullet. A handgun or rifle round consists of a metallic casing, usually brass, a primer, gunpowder, and the bullet which is actually the projectile that is fired out of the gun. A shotgun round, called a shell or shotshell, consists of a plastic case with a metallic base, a primer, gunpowder, a wad, and either a slug or a number of round metallic pellets.

The firearm itself typically consist of several parts:

- Barrel: The steel tube through which bullets are fired.
- Trigger: The lever that fires the gun.
- Chamber: the rear part of the barrel in which the round to be fire is contained.
- Magazine. A container to hold extra ammunition.
- Stock or buttstock. On rifles and shotguns, a wooden or plastic part you put up against your shoulder when firing.
- Grip or pistol grip: On a pistol where you hold the gun in the palm of your hand. Some rifles and

shotguns also have a pistol grip in addition to the buttstock.

- Safety. A mechanical device that prevents the gun from firing.

The basic function of a gun is to propel a projectile down the barrel toward a target. When you pull the trigger a firing pin strikes the primer which sends sparks into the case to ignite the gunpowder. Modern smokeless gunpowder burns very quickly generating gas. This expansion of the gas drives the projectile down and out the barrel.

Because guns are so loud when fired, you should always use hearing protection when you practice. Hearing protection can either be plugs that go into the ear canal or earmuffs that fit over the ears. You may even want to use both in combination for the best protection.

Similarly, you should wear eye protection, either prescription glasses or safety glasses, to protect your eyes from particle discharge.

Handguns

A handgun, either a revolver or a semiautomatic, is a relatively compact firearm that can be carried unobtrusively in a holster, either openly or concealed, in a purse, or even in a pocket. In the military, handguns are usually carried only by officers or as secondary weapons by soldiers. But for law enforcement officers and civilians, a handgun can be a primary weapon.

A handgun is a weapon of convenience. The primary advantage of a handgun is that it can be always with you. If you are not carrying your handgun, if you are storing it for use in an emergency, it almost might as well be a long gun. But a second advantage of a handgun is that it is shorter and easier to manipulate in very close situations such as inside a home. A long gun, particularly one that meets federal regulations, is somewhat cumbersome inside the closed space of a home. However, even within a home, this only applies in certain tight places such as corners. As soon as you have a yard or more clearance the advantage of a handgun is lost.

Handguns come chambered in a variety of calibers ranging from the .22 rimfire to full sized rifle cartridges. The self-defense effectiveness of these calibers varies. Among the most common handgun calibers are the .380 ACP, the 9mm Lugar, the 38 Special, the 357 Magnum, the .40 S&W, and the .45 ACP. There is a great deal of debate about the relative effectiveness of handgun calibers but with modern self-defense hollow point ammunition the differences are insignificant. Hollow point ammunition is designed to expand upon impact and is thus more likely to stop an assailant and less likely to pass through him and hit an innocent bystander.

Some handguns are designed to be carried concealed and, thus, are smaller. Typically, they use smaller cartridges such as the .380 ACP or 9mm Lugar. Smaller handgun calibers, such as 25ACP, 32 Short/Long, and 32 ACP may be better suited to smaller or weaker hands or to better concealment. Larger handguns are primarily designed to be carried openly in a belt holster.

There is controversy about the relative advantages of revolvers vs. semiautomatics. A revolver, typified by the western wheel gun, uses a rotating cylinder to cycle bullets into alignment with the hammer and barrel. A revolver can be either single or double action. A single action revolver requires the hammer to be cocked manually and are no longer used for self-defense, while a double action rotates the cylinder and cocks the hammer with each pull of the trigger. The advantage of the double action revolver, compared to the semiautomatic handgun, is that it is simpler and more reliable and, thus, it can be a better choice for someone who is new to firearms or who does not practice regularly. The disadvantage of double action revolvers is that they have a much longer and heavier trigger pull that women especially find challenging.

A semiautomatic handgun, on the other hand, stores ammunition in a magazine and automatically feeds new rounds into the chamber after each round is fired. Semiautomatics typically hold more ammunition than a revolver and they can be more easily reloaded by simply dropping the empty magazine and inserting a fresh loaded one. More ammunition is better because moving targets are hard to hit, even fatal wounds don't always stop an assailant, and there may be multiple assailants.

The maximum practical effective range of a handgun is approximately 15 yards, depending on your training and skill. Although the energy of a handgun cartridge declines over distance, the main limitation is accuracy. But since most self-defense uses of a handgun occur at two to three yards and few beyond seven yards the inherent accuracy limitations of the handgun are usually not a factor.

However, aiming a handgun is more difficult than a shotgun or a rifle. Steadiness of the handgun rests entirely in your hands since it does not otherwise contact your body. To aim a handgun, grip it with both hands, your weak hand partially covering your strong hand (though it is possible to aim it with only your strong hand or even your weak hand) and align the sights on the target. For the vast majority of handguns this means aligning the front and rear sights (iron sights) on the target. Because your eyes can only focus on one of the three depths, the rear sight, the front sight, and the target, it is recommended that you focus on your front sight. Other sighting devices include a red dot (reflex or holographic) or a laser pointer. Iron sights may feature a glow-in-the-dark paint to make them easier to see in the dark. In addition, many modern handguns are designed to accept a small flashlight attachment that can be used for identifying your target at night.

Training and practice are especially important with handguns. You will need to master the trigger pull, mastering the ability to pull the trigger slowly so that you are (somewhat) surprised when it fires. You will also need to master the recoil of the handgun so that you can quickly realign the sights on the target after each shot. All of this requires live firing at a range, either indoor or outdoor. First get proficient at precision fire. Then learn to fire faster without losing aim. Aiming is necessary for practice but in the heat of a self-defense situation you are more likely to rely on muscle memory and point shooting.

When you return from the range you will need to clean your handgun. This entails brushing the bore clear of residual copper, lead, and burnt gunpowder and oiling or greasing metal parts that are prone to rust during storage.

In addition to live fire at the range you should also practice holster draw and dry fire at home. Dry fire means practicing without live ammunition. Needless to say, the first step in dry fire practice is to remove all live ammunition from the vicinity. Make doubly sure that there is no live ammunition in the handgun or magazines, and none in the vicinity where you will practice. Practice safely drawing your handgun from your holster, putting the sights on a target, and squeezing the trigger. Because there is no noise or recoil, dry firing helps you to reduce the flinch that naturally develops from live firing. Although some guns can be safely dry fired on an empty chamber, others may be damaged. Dummy ammunition solves this problem and allows you to practice loading, unloading, reloading, and clearing malfunctions. There are laser devices that allow you to practice your aiming at home.

Virtually anywhere that a handgun can be legally owned and possessed, a handgun can be legally used for self-defense in the home or place of business. Generally speaking, you are legally permitted to carry a handgun inside your home or business, either concealed or openly, at any time. But laws on carrying a firearm outside your home or business vary widely. Some jurisdictions allow the concealed carrying of a handgun with, or in some cases without, a permit while others make it virtually impossible to do so.

Holsters are designed to attach to a belt or to fit into clothing or a purse and to safely and securely hold a handgun. Some holsters are designed for open carry while others are designed for concealed carry. Concealed carry holsters may be designed to fit outside or inside the waistband. The advantage of inside the waistband holsters is that they are easier to conceal but they are less comfortable.

You should have a holster for your handgun that fits your situation. Even if you don't plan to carry your handgun except in an emergency it is still prudent to have a holster. You may need to do something with your hands, such as dialing for emergency on a phone, and a holster is the safest place for your gun when it is not in your hands or in a safe. When carried in a holster, a handgun can be either loaded or unloaded. The Israeli practice is to have a magazine loaded but to leave the chamber empty. This, then, requires the slide to be racked after drawing from the holster. But most people holster a handgun with a round chambered to avoid that extra step.

On the one hand, to be useful for self-defense a handgun needs to be loaded, or at least the magazine needs to be easily accessible. On the other hand, you don't want to give others, especially children, access to your handgun. A gun safe, which is a secure place to keep your guns, is not easily accessed in an emergency. A minivault, a safe big enough to hold one handgun but small enough to fit in almost any location, is a good compromise. It can be opened with a key, with the press of buttons in a set sequence, or with your fingerprint, though one with a purely mechanical mechanism is more reliable. The handgun can be stored loaded but secured in the minivault. It is possible to quickly load a magazine and load a round in the chamber, but this adds complexity at a time when you are likely to be under tremendous stress.

There are two basic techniques to fire in self-defense against an assailant. The failure drill, also called the Mozambique drill, is to fire two rounds into the body, also known as a double-tap, and then one to the head. A head shot is more difficult but is more likely to lead to immediate incapacitation. The other method is the zipper up in which the handgun is fired as it is drawn in sequence at the assailant's hip, sternum, chest, and finally the head. Once the most dangerous assailant has been incapacitated, you should glance around for others who might be posing a threat before holstering your handgun.

Shotguns

A shotgun is a long gun designed to fire multiple pellets from a single shell, although it can also fire a single slug. Shotguns come in a variety of gauges, but the 12-gauge shotgun is the one most commonly used for self-defense.

The maximum practical effective range of a shotgun is approximately 50 yards, well beyond the typical distance for self-defense situations, especially in the home.

To fire a shotgun, press the butt against your strong shoulder and use your weak hand to support the barrel while your strong hand grips near the action and pulls the trigger. This provides more points of contact and control and improves aiming and recoil management over the handgun.

Shotgun ammunition is classified by the size of the pellets, called shot. Most shot sizes are designed for hunting birds or other small animals. The most common choice for self-defense is 00 buck shot which was originally designed for hunting deer. Each of the approximately nine pellets of 00 buck shot are a third of an inch in diameter, similar to a small handgun bullet. So, in effect, firing a shotgun is like firing nine bullets from a handgun at one time. A shotgun slug, on the other hand, is a single projectile and can be used for precise fire like a rifle.

The traditional hunting shotgun is a double barrel, break action. Each barrel is loaded separately, and both barrels can be fired in succession before the shotgun needs to be reloaded.

A pump (slide) action requires the slide to be articulated after each shot to load a new round. A semiautomatic shotgun, like a semiautomatic handgun, automatically loads a new round after each shot. Both usually have a tube magazine running under the barrel that can hold up to eight shells, depending on the magazine length and the shell length, though some modern shotguns have detachable magazines.

Contrary to popular belief, a shotgun must be aimed. At typical distances inside a home, the shot does not spread out significantly. Shotguns sights usually consist of a single bead at the muzzle of the gun, though some feature front and rear sights. For improved accuracy and quicker acquisition, a red dot sight (reflex or holographic) can be used. A red dot sight is a non-magnifying sight that projects a dot, usually red or green, onto a lens. It does not require the alignment of a front and rear sight; it is sufficient to place the dot on the target to aim the firearm.

A shotgun used for home defense should include an attached flashlight. There are many brands of flashlights, such as Surefire, that are specifically designed to withstand the shock of firing.

Just as a handgun needs a holster, a self-defense shotgun needs a sling. A sling is used to free the hands while carrying the shotgun, but it can also be used to retain control of the shotgun should an assailant try to grab it from you.

There is controversy about the relative merits of the shotgun vs. the handgun for self-defense and each have their strengths and weaknesses. The shotgun is a devastatingly effective self-defense weapon when firing 00 buck shot. And it is much easier to aim and more accurate with more points of contact than a handgun.

But the recoil and concussion of a shotgun are much greater, though because the shotgun is fired from the shoulder, much of the recoil can be absorbed into the body. A shotgun has a lower ammunition capacity than a handgun or a carbine. And due to the length of the barrel, usually a minimum length of 18 inches, a shotgun can be difficult to use in the confined space of a home. But a bullpup design, one that places the action behind the trigger, can shorten the shotgun while remaining within legal constraints.

Carbines

A carbine is short rifle usually firing an intermediate rifle or pistol cartridge. The maximum practical effective range of a carbine is up to 250 yards or more, far beyond the typical distance for self-defense situations.

Like a shotgun, a carbine is fired by pressing the butt of the carbine to your strong shoulder and using your weak hand to support the barrel while your strong hand grips near the action and pulls the trigger.

The modern carbine rifle is usually a derivative of a military assault rifle, and popularly referred to as an 'assault weapon.' But whereas a military assault rifle is selective fire, meaning that you can select between semiautomatic and fully automatic fire, the carbines that civilians typically own are semiautomatic only with one round fired with each pull of the trigger. However, it is like the military assault rifle in every other way including a detachable box magazine that can hold 20 to 30 rounds or more.

The two most common carbines are the AR-15, which fires a .223 Remington or 5.56x45mm NATO cartridge and the AK-47 which fires the 7.62×39mm Russian cartridge. Other carbines suitable for self-defense include pistol caliber carbines and lever action rifles like the Winchester and Marlin that fire revolver cartridges like the 357 magnum, 44 magnum, or 45 Colt. Other rifles are generally not suitable for self-defense.

The bullpup is a recent design innovation that puts the action behind the trigger and shortens the overall length of the carbine without sacrificing the length of the barrel. Many Western European nations field bullpup carbines in their armed forces including the Austrian Steyr AUG, the French FAMAS, and the British L85/86.

United States federal law regulates the minimum length of a rifle, but the AR-15 receiver can be used as the basis of a pistol configuration with a shorter barrel. This configuration makes it easier to maneuver in the tight spaces of a home. To qualify as a pistol, it must not have a shoulder stock. However, due to a quirk in regulations, it can have a so-called 'brace' which is effectively a short stock that can be shouldered.

Traditional carbine sights are front and rear iron sights. To aim the rifle, the front and rear sights are aligned on the target. A red dot (reflex or holographic) is a better option for self-defense. It is aimed simply by placing the dot on the target. As with hunting rifles, magnified sights are available, but they are not suited to the typical ranges of self-defense because it is harder to find the target in a magnified scope. However, a recent innovation is the low power variable optic scope that ranges from 1x and up. When set at 1x it functions essentially as a red dot sight.

Like the shotgun, the carbine needs a sling. Minimally, you need a sling that will allow you to carry the carbine on your shoulder to free your hands. The ideal sling is one that will let the rifle rest in front of you or to your side from where it can be easily presented. Additionally, a sling serves to retain the rifle in case an assailant tries to take it from you.

Like a shotgun, a carbine is easier to aim and control than a handgun because it has more points of contact. But it is more accurate and precise and has less recoil than a shotgun. It has a higher ammunition capacity than either a shotgun or a handgun, usually 20 to 30 rounds. And the intermediate rifle cartridge is more effective than either handgun or shotgun ammunition. However, like shotguns, carbines are longer than a handgun and have a powerful concussion. Overall, a carbine is probably the type of gun most easily mastered for self-defense.

Although some people prefer a carbine for self-defense in the home, it is especially well suited to defending a wider space such as a neighborhood or a commercial building from larger threats such as riots. During the 1992 rioting in Los Angeles, for example, shop owners in Koreatown, upon learning that they were the next target, armed themselves with firearms, including AR-15s and AK-47s, and took up overwatch positions on the rooftops of their stores from which they had a wide field of fire into the surrounding parking lots.

On October 7, 2023, Hamas fighters invaded Israel from Gaza attacking military outposts and villages alike. The military was caught entirely off guard, and it was hours before they were able to respond. In the meantime, Israeli citizens were on their own. Due to severe restrictions on private ownership of firearms, few Israeli citizens were prepared and Hamas fighters were able to kill almost with impunity. But in a few cases, Israeli citizens were able to take up arms, typically M16s and AR15 carbines, and defend their homes and villages. Since the attack Israel has begun relaxing its restrictions on private ownership of arms.

Home

Your home is your castle. And so most legal jurisdictions allow wide latitude in the exercise of self-defense in our own homes and the carrying and use of weapons for that purpose. They rightly presume that a criminal invading an occupied dwelling intends bodily harm.

While burglary is not ordinarily a matter of self-defense since, by definition, it occurs when you are away from home, nevertheless, it sometimes happens that burglars enter a home expecting it to be unoccupied and, when it is not, escalating to crimes of violence.

To reduce the risk of burglary as well as the risk of home invasion, you should harden your home and develop a plan to turn the tables on an intruder.

Lock your doors and windows. They will discourage a casual thief but also delay a determined assailant. You can strengthen your doors by replacing the short screws in the hinges and strike plates with longer screws.

Get a dog, one that will bark loudly at strangers. But even a sign warning of a dangerous dog can serve as a deterrent.

Get and use a home alarm system. Light up the exterior of your home with motion activated lights and cameras.

Install a camera at your front door, or at least peep hole. When someone knocks at your front door, look to see who is there before opening it. Be wary about opening the door to a stranger you were not expecting, especially at night. Be wary of someone asking for refuge or assistance or to use your phone at night. When in doubt, call 911 for them. If you hear a knock at night, go to the door armed and prepared.

If a burglar comes to your home and rings the door, merely answering should be sufficient to discourage him. But that will not be the case for home invaders or someone of diminished mental capacity due to drugs or alcohol. When you answer the door, especially when you are not expecting someone, don't recognize who is at the door (through the camera or peep hole), or it is an unusual time (such as night), answer cautiously placing yourself at least partially behind the door. You can use your foot as a door stop if he suddenly tries to force his way in and that will give you time to respond. You can slam the door back against them and, perhaps, get it closed and locked again. Call out to others in the house to warn them and to get them to call for the police. If you are armed, this might be the time to draw or even shoot, if necessary, to prevent forced entry. The door becomes a fatal funnel for the assailants. But don't shoot blindly through a closed door.

If someone breaks into your home, then your best option is to barricade yourself in your bedroom with a weapon and call 911. Depending upon the intent of the home invaders, they may steal property and then leave. But if, in the worst case, they break down your bedroom door, you will be prepared to ambush them, and they will be at a great tactical disadvantage.

However, that is not always practical. There are two reasons why you might need to leave the security of your bedroom. First, if there are other family members scattered around the home you may need to take up a position in a hallway that allows you to protect your loved ones without unduly exposing yourself to the home invaders. In the worst case, you may need to fight your way to them.

But it may happen that you hear a bump in the night, or a dog barking, that doesn't make sense to you. If you are concerned about it, call 911. But calling 911 every time you hear a bump in the night may lead the police to downgrade their response to your home or bill you for the call. On the other hand, you don't want to ignore the mysterious sound and go back to sleep. It is in these situations that you must take up the dangerous task of clearing your own home.

Clearing your home means checking every room for someone who may be hiding. This is a very dangerous undertaking if there is, in fact, a home invader because now you risk being ambushed in your own home. So, clear your home with a flashlight and a weapon at the ready. A handgun is particularly useful for room clearing because it is smaller and more easily maneuvered in tight spaces such as a corridor, but a bullpup shotgun or carbine with a light can also be a good choice. Even a baseball bat is better than nothing.

Traveling

Your options for self-defense are more limited when you are traveling outside the home. Not only do you face more stringent laws on the bearing of arms and their use in self-defense, but you are also generally in unfamiliar places dealing with unfamiliar people. Especially when you travel internationally, you are in a new and different environment, perhaps facing unfamiliar social customs, and unfamiliar laws, maybe even unfamiliar languages.

Awareness and alertness are especially crucial when traveling. When you are on foot walk briskly and confidently with your chin up. Be alert, keep your eyes moving. Check everything around you. Be especially wary of attacks from behind, Be careful of blind corners. Always keep at least one hand free.

Try not to stand out but know that you will anyway. Keep to a group if possible. And don't carry a lot of cash.

Public transportation poses special risks due to crowding and your inability to leave. Pickpockets may stealthily rob you. But more serious crimes, such as assault or rape, have occurred at night when there are fewer bystanders or even in broad daylight when brazen gangs simply ignore bystanders altogether.

When you are driving you are at risk of carjacking. A carjacker might attack when you are trapped at a signal light or they may use a ploy such as faking an accident, either with another car or as a pedestrian. Clever carjackers have been known to drive nails into tires when you park that leave you with a flat on the road. Then they come to your 'rescue.'

Lock your car doors and keep your windows rolled up. When you stop in a line of cars, keep the rear tires of the car in front of you in view to give you space to move out of your lane if you need to escape. Keep at least a quarter tank of gas to avoid running empty. Be a defensive and courteous driver and avoid road rage situations.

Never allow yourself to be abducted to a secluded place. If you are driving slam on the brakes or accelerate into a parked car or a building.

Women, but also men, have been victims of rape or robbery from spiked drinks. Be very careful what you ingest when you are with strangers or anyone that you don't trust.

If you are staying at a hotel or a motel don't let anyone know you are staying alone, check your room before going to sleep and use a door alarm.

Offices create a difficult security challenge. Most companies prohibit self-defense weapons at the office. This is less of an issue when the office provides security or during the day when there many others around. But if you work late at night after everyone else has gone home you may be putting yourself at risk.

Offices are known 'gun free zones' and can attract terrorists bent on creating mass casualties. If you ever find yourself in an active shooter situation, try first to escape from the location. If that is not possible then barricade yourself in a room by locking and/or barring the door with furniture and find an escape route. If that is not possible, or if the assailant breaks in, your last option is to fight.

Children

Children face unique threats. Young children, especially, are at a greater disadvantage to those who would prey on them, both mentally and physically. We simply cannot rely on our children to be aware and alert much less to be prepared to defend themselves. Children, and even young adults, tend to be particularly naïve regarding the dangers of the world around them.

Children are not typically the target of property crimes, but they may be targeted for a cellphone or laptop.

Abduction is a parent's greatest fear. On the one hand, abduction statistics are somewhat misleading. The vast majority of child abductions are by a parent in a child custody dispute; the child is not a target, just a pawn. But, on the other hand, abductions by strangers, rare as they are, are far more heinous in nature including molestation, rape, torture, and murder.

It is your responsibility, as a parent, to teach your children personal safety. This is a skill that will follow them throughout their lives. And, fortunately, it is possible to do this without terrorizing them or filing them with a fear of the world. Teach your children confidence, not paranoia.

One of the most common lessons that children are taught is, in fact, bad advice. We have all heard the admonishment to "never talk to strangers." This can lead to a general distrust of adults that the child does not know, even police officers. But most adults are not a threat to your children. Instead, you should encourage your child to seek help from an adult whenever he needs it and to be wary only of adults that pay undue attention to them. It is far safer for your child to randomly choose someone to request help from than to accept help from an adult who has initiated an interaction with them.

Your children's best defense against predators is:

- A sense of their own rights.
- A solid understanding of safety rules.
- The ability to assess situations.
- Knowing when and how to seek help.
- Confidence they will be believed.

Use role playing to teach your children about personal safety. Act out situations that might be dangerous and test them on their response.

Be a good example for your children by demonstrating awareness and alertness. Encourage them to be aware and alert with you by inviting them into the game. Award points for identifying suspicious people or dangerous situations.

Learn to Say 'No!'

We want our children to respect and obey us as well as their teachers and other figures of authority, but they should understand that they can say 'no' to strangers and even draw boundaries with authorities. They don't always have to be polite.

Identify Threatening Strangers

Children are not good judges of who is a threat. They tend to focus on scary looks or unusual behavior as modeled in film and television. Instead, encourage your child to be suspicious of unwarranted attention from strangers but to be confident about asking for help from strangers.

Beware of Tricky People

You may have heard the admonition, "don't take candy from strangers," but this is just the tip of the iceberg. Common lines that predators use include, "can you help me find my puppy", "come with me, I know your parents", "I can make you famous", and "don't tell anyone or else…". Teach your child to leave a suspicious conversation.

Walk Tall

Like adults, children who exude confidence and alertness are less likely to attract predatory attention. Teach your child to pay attention to his surroundings anytime he is not with a trusted adult.

Don't Answer the Door

Children should be discouraged from answering the door, especially when they are left home alone. Teach your child to let an adult answer the door, even when they are expecting friends.

Always Have a Buddy

Predators prefer easy victims and even one friend can complicate things. Teach your children that anytime they are not with a trustworthy adult they should be with a friend.

Learn a Codeword

There are many situations where a code word can be useful. For example, if you ask a friend to pick up your child from school, a code word will let the child know that this adult is trusted by you.

Yell! Yell! Yell!

Children don't need a lot of encouragement to yell but they don't always know when and what to yell to protect themselves. Teach your child to yell for help whenever he feels threatened by a stranger.

Break Away

While a child is not going to prevail over an adult in a fight, he can often wring himself free. Teach your children how to break away, evade, and escape either to safety or to a hiding place.

Run Like the Wind

Many abductions are thwarted when the child simply escapes to safety. Although adults are usually faster, a predator might not be willing to risk exposing himself by giving chase. Encourage your child to run as fast as he can when he feels threatened by an adult.

Mobs

Violent mobs pose a particularly difficult challenge for self-defense. By far the best way to avoid the risk of a mob attack is to simply avoid being at the time and place where violent mobs are most likely to gather. That includes, for example, political demonstrations and riots. Or, if you suspect that a peaceful demonstration is becoming a riot then leave as quickly as you can by the safest means available.

Violent mobs pose four unique risks: First, the nature of a mob is that it confers anonymity on those who act within it. This permits people to act out in ways that they would never otherwise dare. In fact, people tend to behave very differently when they are swept up in the emotions of a mob. Otherwise, peaceful people can become fanatically violent when they feel they have the approval and urging of those around them.

Second, mobs have numbers on their side. Armed and trained police officers will often choose to beat a hasty retreat when faced with a violent mob. Even formations of police officers must be careful not to allow themselves to be isolated and overwhelmed.

Third, mobs can attack from many directions and therefore can attack you from behind. While one person is taunting you to your face another may sneak up and punch you in the back of the head.

Fourth, very often the goal of a mob is to hurt someone, perhaps you. Ordinary criminals are usually content to steal from you. When you surrender your wallet or purse, they will seek to escape. Not so a mob. When a mob turns violent, they are looking for someone to injure, perhaps even to kill. This is particularly so at a political demonstration where two opposing factions face off; even if you are not a party to either faction you can easily be mistaken for it.

If you cannot avoid or escape a confrontation with a mob, then the next best thing is to be prepared.

There are four scenarios to consider depending on where you are when you encounter the mob: at home, in your car, at a restaurant, on foot.

Defending yourself against a mob in your own home is arguably the easiest. There, at least, you are in a familiar place, you have access to whatever weapons you keep at home, and you enjoy the greatest legal protections and, thus, liberties. While a mob at your home might be frightening, it only poses a danger if it tries either to break into your home, set fire to your home, or shoot into it with firearms. In these cases, you will be justified in using deadly force to repel the mob. Otherwise, if the mob is merely trashing your yard or throwing bricks and rocks into your windows, you will need to use your own judgement as to what level of force is appropriate, if any.

Recently, mobs have started blocking streets and even freeways as a form of protest. Mostly they have limited themselves to delaying, harassing drivers passing through but, in some cases, drivers have been pulled from their vehicles and beaten and in other cases cars have been damaged. And even when the intent is just to harass, some drivers have been forced to perform humiliating acts of allegiance to the mob in order to pass through the roadblock.

You might be driving along minding your own business and suddenly you realize that you have driven into a political protest, and you are now surrounded by an angry mob, mostly college students and young adults.

If you drive into a mob, stay in your car. Roll up the windows and lock the doors. If you see a mob ahead of you, reverse or turn around and get out of there.

Regardless of what weapons you might have, your automobile, itself, is a deadly weapon. That means that so long as you are not in imminent danger, you must use great caution while driving through a mob. This might require you to drive at a crawl or even to come to a complete stop even if the mob is damaging your automobile.

The situation changes, entirely, the moment that anyone in the mob endangers you and the other passengers in the automobile. If, for example, someone in the mob begins breaking through a window or points a gun at you or otherwise presents an imminent threat to your safety, you can use deadly force to escape including, if necessary, driving over anyone in your path.

While a firearm is handy in a situation like this, it is important to remember that your vehicle is, itself, a deadly weapon that can be used judiciously.

Another mob tactic is to invade a restaurant to harass diners. These are not typically violent attacks but only threats of violence to force allegiance to the cause they are advocating. Usually, you will be safe if you comply with their demands.

Easily the most dangerous situation is when you face a mob on foot.

Mobs may be armed in a variety of ways. Sheer numbers are dangerous enough that most people participating in a mob may not be armed. Those who are will typically be armed with blunt weapons such as bats and bricks. You may occasionally encounter more exotic weapons such as slingshots or Molotov cocktails. Because mobs usually try to avoid provoking an armed response by police, they rarely carry firearms.

Being armed for self-defense, ideally with a concealed firearm or, at a minimum, a knife, may give you a superiority in arms.

There are some basic tactics that you can employ to reduce the advantages of a mob.

You cannot hope to match the mob in numbers but, due to simple geometry, the greater your own numbers the smaller the mob's advantage. Imagine drawing your group into a circle. One or two people will easily be surrounded by four or more. But as the number in your group grows, the circumference of the circle grows and it begins to approximate a line in which the mob has no numerical advantage, no matter how many others may be stacked up behind those in front, the immediate numerical advantage is much less.

Finally, assuming that you are roughly equally armed and that you have at least four or five people in your group capable of acting in self-defense, you can further improve your chances of survival and protecting the weaker among the group by utilizing some simple tactics. Form a circle with the weakest members of the group (children, the elderly, and women) inside and those capable of acting in self-defense on the outside. Those on the outside should face the mob, even if you are on the move. Each person on the outside will face one or two (hopefully not more) members of the mob at a time. There is not much point in knocking out a member of a mob, he would just be replaced by another. Instead, aim to keep the mob at bay and protect yourself and those inside the circle from harm by simply pushing back.

In any situation where you are facing a mob, look for the instigator. Strike the shepherd and the sheep will scatter.

Practice and Training

You can only learn so much about self-defense from reading books and articles. Owning weapons is of no use if you don't know how to use them. When you are in a dangerous situation, your adrenaline will impair your thinking and you will tend to act out your training, if any, or to shut down entirely. Fortunately, there are many things you can do to prepare to defend yourself and your family.

The internet has made information ubiquitous, and this includes not only written text but also videos on platforms like YouTube and educational platforms such as Udemy.

But eventually you will need to begin practicing what you learn. Practice helps you develop what is called 'muscle memory,' habitual and automatic behaviors that you can perform in a self-defense situation with little or no conscious effort. Timed exercise and competition can help inoculate you against the stress of violence.

To be certain that you are developing the right self-defense habits you should consider investing in formal training. Otherwise, you risk learning the wrong behaviors. Self-defense training can range from unarmed martial arts to weapons training to intensive courses that simulate self-defense situations.

Training in martial arts, for example, typically incudes practice of basic self-defense forms, punching, kicking, and blocking, as well as sparring. Training helps you test and develop your endurance. Sparring, especially, will help you develop mental toughness and resilience; you will learn that just because you are hit doesn't mean you are out of the fight.

But there are practical limitations to every training option. Martial arts pursue several goals including exercise, sport, and art form. Self-defense is not the only or, in most cases, even the main goal. Training without an opponent becomes stylized and sparring with an opponent must be restricted to prevent injury. Nevertheless, martial arts allow you to rehearse different situations both mentally and physically.

Training with firearms can begin with learning how to manipulate, load, aim, and shoot at a public range and then graduate to drawing from a holster, to firing with a high heart rate and under duress, to utilize cover (protection) and concealment (hiding), to shoot from different positions and on the move at multiple targets, and to communicate with others.

Don't forget to train and practice with your family to share what you have learned. Develop emergency plans that include dealing with the most common threats to personal safety and practice these with your family. For example, sound the alarm in the middle of the night to test your family's response.

Aftermath

If you are ever involved in defending yourself, regardless of the outcome, your first step should be to call, or have someone else call, for emergency services if only to position yourself in the minds of the authorities as the victim. Let the person answering know if someone has been injured or killed and if the threat has been resolved or if it is still ongoing.

Once emergency personnel arrive, usually the police and perhaps a paramedic or ambulance, let them take control of the situation. Until then, you should maintain control to the best of your ability including, for example, keeping an assailant at gunpoint and rendering aid to anyone injured, if that does not endanger yourself or others.

If, while defending yourself, you injured or killed your assailant, especially if this was with a firearm, you should expect to be intensely questioned by police who will seek to establish the circumstances. Depending on the jurisdiction and the circumstances, the police may determine on the spot whether to arrest you on suspicion of manslaughter. Or your arrest may be routine and the decision to prosecute left to a city attorney.

If possible, you do not want to have the gun in your hand when the police arrive. Holster it or set it down somewhere nearby. (But don't try to hide it.)

When the police ask you what happened, you should reply, simply: "I was in fear for my life (or the lives of others). I want to speak to an attorney." The United States Supreme Court long ago decided in the Miranda case that the Fifth Amendment to the United States Constitution affords you the right to remain silent and to an attorney but that anything you say can be used against you in a court of law. Even in foreign jurisdictions where this constitutional protection is not afforded, you are almost certainly better off saying as little as possible. In other words, just shut up and let the legal system resolve the matter, hopefully in your favor.

But self-defense instructor Massad Ayoob notes that the deck may already be stacked against you and suggests making a good, polite first impression with responding officers:

1. Say "I was attacked" or "I was threatened with attack" or "I felt like my life was threatened."

2. Say "I will testify in court against him, I will press charges and sign a complaint."
3. Point out the physical evidence, if any, of the assailant's criminal intent.
4. Identify any witnesses to the assault.
5. Finally, say "Officer, you will have my full cooperation after I've spoken with legal counsel."

While self-defense is a natural right, it does not enjoy universal support. In the aftermath of exercising self-defense there is a good chance that you will encounter those who pay lip service to the right but oppose any practical application of it. Their attitude was eloquently captured by Alexander Murdoch Gow who, in 1873 wrote in *Good Morals and Gentle Manners*:

> Next to the evil influence of an unruly tongue in stimulating murderous designs is the habit of carrying deadly weapons and practicing with them. It is not an uncommon thing for young men and boys to carry knives, pistols, slingshots, and other death-dealing instruments. What a fearful account must they render who cherish hatred, who encourage strife and coolly deliberate upon

> the taking of life by practicing for the consummation of the mischief.
> It is not strange that a savage, who does not recognize and is not governed by high moral obligations, should continually practice with his weapons, in order to preserve his life and punish his enemy. Nor is it surprising that a soldier, whose trade is to kill, should be so employed; but for a citizen in an enlightened land, under ordinary circumstances, to carry concealed weapons, is a confession of weakness and cowardice of which no truly brave person can be guilty. It leads also to the violation of the civil law, for when a person has prepared himself for a deadly conflict, when his heart is set on it, when the murderous intent is in his words and actions, it is but a short step to the taking of life.

You should assume you are going to be prosecuted for exercising your right of self-defense no matter how innocently you may have acted. By assuming the worst, you will be better prepared for the ensuing inquiry.

Conclusion

Preparing to exercise your right to self-defense begins with your mental attitude. Think through as many scenarios as you can imagine and how you would respond in each. Work out your opinion on using deadly force to protect your life or the lives of those you love.

Next, evaluate your life and where you are at the greatest risk from crime. Is your neighborhood unsafe? Do you travel to work on public transit or is your office located in a dangerous part of town? Crime statistics can tell you a lot about the risks in your area and where you should be concentrating your self-defense preparations. Online forums can be a great place to learn informally from others with a similar interest.

Research what weapons you are allowed to own and carry in your legal jurisdiction. Remember, what is legal at home may not be legal at work or in public. Request a ride-along with your local police department or sheriff to learn more about what is going on in your neighborhood, how police respond to emergencies, and about local policies and informal recommendations.

Train in self-defense and practice what you learn. If circumstances permit, acquire a firearm and become proficient with it.

There are many good reasons to be physically fit but effectiveness with natural weapons in self-defense is one more. Martial arts training is a great way to improve your self-defense skills while improving your physical fitness.

Remember, a gentleman always goes armed. If not a concealed firearm, then a folding knife or whatever your legal jurisdiction permits.

Bibliography

The material presented in this book was drawn, in part, from the following sources.

Wikipedia.org

The Dirty Dozen: 12 Nasty Fighting Techniques for Any Self-Defense Situation by Larry Jordan, 2002.

Principles of Personal Defense by Jeff Cooper, 1989.

Unleashing the Warrior Within: Using the 7 Principles of Combat to Achieve Your Goals by Richard J. Machowicz, 2000.

Waking the Tiger Within: How to Be Safe from Crime by Scott Flint, 2001.

For further reading, visit: www.conquistador.org > Academy > Conflict > Self Defense.

Edward A. Ipser, Jr., PhD

Edward A. Ipser, Jr., PhD is President of IpserLab LLC, a startup foundry and Managing Partner at Coactify Inc., an innovation consultancy as well as adjunct faculty at TCU. He has founded, cofounded, joined, and advised many startups. As the founder of MarketFusion he raised $23M in venture capital. He also founded the Silicon Valley Association of Software Entrepreneurs and the Fort Worth Innovators Forum. He graduated from TCU with majors in Math, Psychology, and Economics and received a PhD in computer science from USC.

Conquering Life Series

Conquering Coaching:
How to Bring Out the Best in Others

Conquering Creativity:
How to Discover Original Ideas

Conquering Critical Thinking:
How to Form Sound Judgment

Conquering Innovation:
How to Create Valuable Solutions

Conquering Marketing:
How to Promote Your Products and Services

Conquering Morality:
How to Know Right from Wrong

Conquering Rhetoric:
How to Influence Through Your Words

Conquering Sales:
How to Sell Products and Services

Conquering Self Defense:
How to Protect Yourself and Those You Love

Conquering Startups:
How to Turn an Idea into a Successful Business

Available at
www.conquistador.org

Sociedad de los Conquistadores

Conquistador in Spanish means 'conqueror,' one who overcomes or surmounts. Historically, the word refers to Spaniards without inheritance who led expeditions to the newly discovered Americas in search of fortune and fame. The most successful conquistador was Hernán Cortés, who in 1521, with some 800 men and 16 horses, armed with swords, lances, crossbows, and a few small cannon, and with the aid of native American allies, overran the mighty Aztec Empire in Mexico. Francisco Pizarro similarly conquered the Inca Empire in Peru.

Sociedad de los Conquistadores, Society of the Conquistadors, is dedicated to celebrating human exploration and conquest in business and life. While postmodern egalitarian ideals have bred a cult of mediocrity and a repression of individual ambition, today's Conquistadors are unashamedly dedicated to personal excellence, achievement, and success, to our own private *reconquista*.

Made in the USA
Monee, IL
04 November 2026

67384590R00066